Mike's Civil PE Exam Guide

Morning Session

Mike Hansen, P.E.

Feel confident on exam day because you know you are prepared!

www.PEexamguides.com

ERRATA

This is the first version of Mike's PE Exam Guide to be released. Perfection was attempted, but we are all flawed. Visit www.PEeexamguides.com to report or view corrections that have been made to this material. Your help is appreciated.

Mike's Civil PE Exam Guide: Morning Session

First Edition

Copyright © 2010 by Mike Hansen

Cover design by Hansen Structural Design by Mike Hansen

Printed in the United States of America

First Printing: July 2010

Updated: October 7, 2010

Mike Hansen, P.E.
PEexamhelp@gmail.com
www.PEexamguides.com

Contents

About the Author

I wrote this book as a 26-year-old engineer who just passed the Civil PE exam with an emphasis on hydrology/environmental for the afternoon section. I graduated from Arizona State University, 2006 and got my Master's in Business Administration from the John Sterling School of Business in 2009. I currently work as assistant construction manager for Wood, Patel & Associates in Phoenix.

Preface

Many books share knowledgeable information about the NCEES PE exam for civil engineers. The favored book amongst those studying for the PE is the Civil Engineering Reference Manual (The CERM) by Michael R Lindeburg. The CERM closely resembles the clunky, data intensive books we have seen throughout school. It consists of very valuable information amongst lots of theory, history, and other useless information that will not help you pass the test. The CERM however remains the best one stop shop for information on the test. This guide will aid you step-by-step through forty Civil PE Exam morning style questions. Not only do the example problems show you how to complete the problem, but they also show you why certain formulas are used, and where they can be found in the CERM. You should have your own copy of the CERM and this guide for the test. The goal of this guide is not for you to relearn everything from school, but to get you to pass the PE exam.

I studied for the Civil PE Exam with two friends and our study methods allowed us to all pass the first time (October 2009). Our study methods are reflected in the step-by-step process that are shown in the solutions to each problem. The only book I brought to the exam was the CERM. This guide will help you utilize the CERM to its full potential, as well as give you a firm grasp on how to accurately complete problems within six minutes.

My Issue with other books

We all desperately want to pass the PE exam the first time through. I bought the Practice Problems for the Civil Engineering PE Exam by Michael R. Lindeburg, PE and Six-Minute Solutions for Civil PE Exam Problems by R.Wane Schneiter, PhD, PE, DEE.

Practice Problems for the Civil Engineering PE Exam

I bought this book of practice problems from the same author of the CERM. It was not shocking to find out that the practice problems closely resemble the lengthy chapters found in the CERM. Simply stated the problems are too difficult and too long. Many of the problems in the book give you a time limit of one hour to complete the question. A one-hour long question is ridiculous! You only have an average of six minutes per problem. The book contains a few good examples amongst many terribly in depth questions that you will never see on the exam.

Six-Minute Solutions

Finally, no more hour-long problems to solve. This book makes sense, right? You have an average of six minutes per problem when you take the exam. While the problems in this book are much more realistic than the Lindeburg problems, they were still more difficult than questions on the test. I looked through the book the day following the exam and only saw a couple similar questions. If you desperately need more example problems, even though these are more difficult, they are effective.

Introduction

Test Advice

The Exam

- The questions on the exam are in no particular order. You can expect to see a hydrology problem directly followed by a structural problem, then go right back to hydrology.

- When I took the exam, 1 in 8 of the questions was a theory question. The glossary in the CERM or index can help you quickly reference the topic in question.

- Each question on the exam is its own question. In my experience, Problem 6 was never related to the answer from Problem 5.

- Does your answer make sense? When you complete a problem take a look at the numbers and the question, does your answer make sense? Use your engineering judgment to justify your answer. They will try to trick you with unit conversion, so be careful with units.

- The practice problems throughout this guide are in a format similar to what you will see on the exam. Get familiar with the format so you are comfortable on exam day.

- If you see a problem you are not comfortable with then skip it and move to the next problem. If you have time at the end of the test then go back and try it. If you do not have time to solve it then guess. You are not penalized for guessing.

- The exam will contain problems similar to the problems you will see in this guide. The exam will also contain problems that are not in this guide. Be comfortable knowing that you will see problems that you have not attempted before. Utilize the study methods throughout this guide to help you solve each problem.

- I brought a three-ring binder full of random equations I found along the way, practice problems, and notes. Bind any notes you use to study and clearly organize them for quick reference.

- Tip from a friend (Andrew Jupp, PE) – Make a copy of the index from the CERM and keep it separate. It can be difficult flipping the large book back and forth. This will also let you keep your appendix page open while you flip through the book.

Things to Bring

- Pack your own lunch in a cooler. I saw two people that did not make it back into the testing center in time and they were rejected from the afternoon portion of the exam! I brought an ice chest with Pringles, a ham sandwich, and a Gatorade.

- Dress accordingly. Assume the testing center will be cold and bring a light jacket.

- If you are easily distracted then bring earplugs. There may be nearby construction, noisy neighbors, and/or ringing cell phones that divert your attention.

- Extra batteries for your calculator

- A photo ID

- The admissions slip NCEES or the state board sent you in the mail. Seat number, etc.

- Food and drink are not allowed on top of the desk, but you can keep it on the floor.

- **Bring an approved calculator!!** Check the NCEES site for a list of approved calculators.

- **Leave your cell phone in your vehicle.**

- The testing center will provide mechanical pencils for your use, but it still does not hurt to bring your own as a backup.

- Few extra single dollar bills and change for vending machines.

Optional Items to Bring

- Seat Cushion

- Ruler/Straight Edge

- Box on wheels to carry your book(s)

The Day Before the Exam

- Drive to your test site so you know where it is.

- Organize your material and make sure you have everything ready to go.

- Set your alarm and a backup alarm. Have a parent/friend call you to make sure you are awake.

- Eat a healthy dinner the night before. Protein, carbohydrates, and a vegetable. Avoid spicy food or anything that could give you the runs.

- Glance over all your material and make sure your tabs are clearly labeled.

- Go to bed at a reasonable time. If your test is at 7am, you should be in bed by 8-9pm.

The Day of the Exam

- Wake up early. Leave so you arrive at the test site an hour before the doors open.

- Eat a light breakfast. Do not drink too much water.

- Go to the bathroom 15 minutes before they open the doors.

- Say hi to people you know, or get friendly with your test neighbors. Get comfortable. If you do not like people, then just sit and relax.

- Organize your desk and get ready.

The Morning Session Breakdown

The morning session consists of 40 Multiple-choice questions from Construction (20%), Geotech (20%), Structures (20%), Transportation (20%), Water Resources, and Environmental (20%). In other words, that is eight problems per section.

Study Schedule

Many people begin to study for the exam up to three months in advance. I would recommend starting no later than two months before the exam. Start by gradually introducing yourself to example questions. Do a few example problems on a free night and really attempt to understand the solution. On the weekends, spend a couple of hours with a friend, coworker, or someone who is also taking the test. Work problems out together, challenge each other to solve a question without looking at the solutions page. It should take you no longer than six minutes per problem.

Three weeks prior to the exam, you should be devoting all free time to studying. I recommend at least four hours a day on the weekend. Flip through your CERM and tab any pages you think may be on the exam. Make sure you know how to utilize your tabs during the test.

How to use this book:
STEP 1: Problem Analysis

1. What genre of problem are you looking at?

   a. Hydrology, Traffic, Geotechnical, Structural, or Construction

   b. More specific, do you know what type of problem it is?

      i. Ex] Hydrology – Open Channel Flow

2. Every question asks you to find X, determine what X is from the problem statement.

STEP 2: Reference

1. Flip open your CERM to the appropriate section according to Step 1 by utilizing the tabs you have placed in your book.

   a. Find the formula or equation that applies specifically to the problem.

   b. If you can solve the problem with the equation, do it. Many problems will require 2-steps. You may need Y to solve for X in your equation. Look for another equation that solves for Y so you can solve for X. As you look through the solutions, you will understand what is meant here.

Note: I have over 50 tabs in my CERM and during the exam, and I used them frequently. Color Code each section so you have each section with a different color tab. Hydrology had green tabs only, Structural was blue, etc…

STEP 3-?: Solve

1. The following 40 example problems show you step-by-step the easiest way to complete each problem.

The Burm Question:

My friend and study partner Jason Burm, P.E. was never satisfied with just the answer to a problem. After every question was completed, he would look at it from another angle and say, "What if they asked us about this…" Therefore, I have aptly added a section to some of the problems throughout the guide labeled "*The Burm Question*". Utilize the Burm Question as an opportunity to look at the problem differently and it is possible something similar will even be on the exam.

Set of 40 Example Problems

Problem 1

A retaining wall is supporting a soil with the properties shown in the figure. What is the total active resultant per unit width of wall?

a) 18 kips/ft
b) 21 kips/ft
c) 24 kips/ft
d) 30 kips/ft

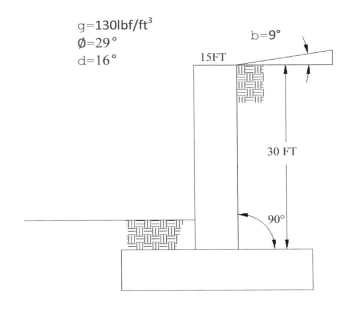

Problem 2

In asphalt paving, a thicker section of pavement (t >4") has many advantageous benefits. Which of the following is not one of the benefits?

a) Lifts can be placed in cooler weather
b) More economical to place one lift versus two
c) It can be easier to reach density requirements
d) Requires less time to setup before drivers can use it

Problem 3

A 6ft x 8ft reinforced concrete box culvert (n=0.013) receives a flow of 215cfs from an irrigation canal. The slope of the culvert is 2%. Assume the culvert does not flow full. What is the flow depth in the box culvert?

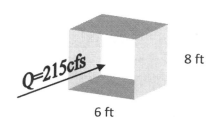

8 ft

6 ft

   a) 2ft
   b) 3ft
   c) 5ft
   d) 6ft

Problem 4

The pump system in the figure is drawing water from the lake and pumping it into the tank. The pump efficiency is 90%, the Darcy friction factor is 0.02, the diameter of the pipe is 4 inches, pipe length is 100 ft, and flow rate is 240 Gallons per minute.

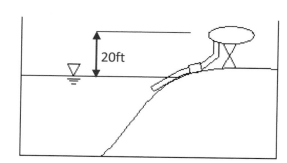

20ft

   a) 0.5 hp
   b) 1.0 hp
   c) 1.5 hp
   d) 2.5 hp

Problem 5

A new roadway is to be constructed and the contractor is preparing grade. According to the stations and cut/fill shown in the table, what is the amount of excavation required?

Station	Fill(+) or Cut (-)
1+00	+50ft^2
1+50	-15 ft^2
2+50	-75 ft^2
4+00	0 ft^2

a) 225 yd^3
b) 343 yd^3
c) 650 yd^3
d) 1025 yd^3

Problem 6

Which of the following statements about open channel cross sections is true?

a) An efficient open channel cross section minimizes flow.

b) The most efficient rectangle cross section has a depth equal to one third the width: $d = \dfrac{w}{3}$

c) A semicircular cross section is the most efficient shape and minimizes construction cost.

d) The most efficient cross section requires the hydraulic radius be maximum and the wetted perimeter minimized.

Problem 7

A sieve analysis was completed on native soil where a homebuilder is proposing a new subdivision. What is the USCS classification for this sample?

a) SW
b) CH
c) MH
d) CL

Sieve Number	Percent Finer
1.5 inch	98
No. 4	85
No. 10	72
No. 20	68
No. 40	62
No. 100	57
No. 200	52

Liquid Limit = 55
Plastic Limit = 27

Problem 8

Which of the shear diagrams most-accurately resembles the beam loading shown in the figure?

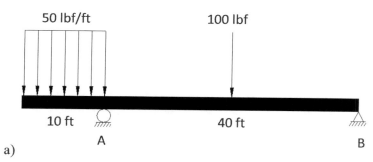

a)

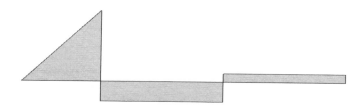

b)

c)

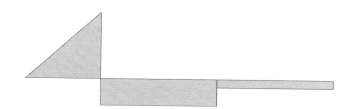

d)

MIKE HANSEN, P.E.

Problem 9

A contractor is looking to build a headwall for a 36" storm drain outlet. How much plywood should he purchase for forming the headwall structure? Assume there is 10% waste over the exact calculation. The headwall is 1ft thick.

a) 89.93 ft^2
b) 98.93 ft^2
c) 179.86 ft^2
d) 193.45 ft^2

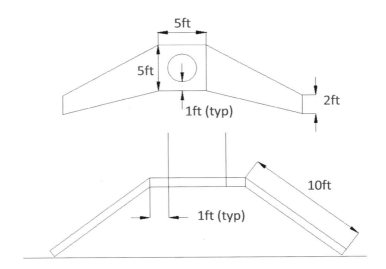

Problem 10

Use the construction schedule in the table to determine the duration of the critical path.

a) 13 days
b) 15 days
c) 17 days
d) 18 days

Activity	Duration (days)	Predecessor
A - Excavate Box Culvert	4	-
B - Excavate Turn Downs	1	A
C - Place Footing Forms	2	A
D - Place Footing Rebar	3	A
E - Pour Footing	1	B,C,D
F - Place Box Culvert Rebar	5	E
G - Set Box Culvert Forms	3	E
H - Pour Box Culvert	1	F,G
I - Backfill Structure	3	H

Problem 11

In the given truss, what force does member BC experience?

a) 11k Compression
b) 5k Compression
c) 1k Tension
d) 11k Tension

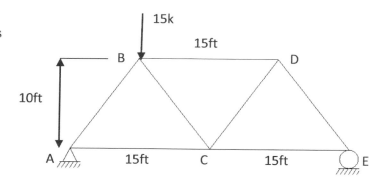

Problem 12

A horizontal curve with a radius of 1500ft has the PI shown in the figure. What is the station of the PT?

$I=14°$

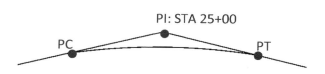

a) 13+41
b) 14+67
c) 25+33
d) 26+83

Problem 13

A sports car is speeding down a straightaway at 80MPH when he notices a stop sign rapidly approaching. Considering his perception reaction time (assume 2.5 seconds), how far will he travel before the car comes to a stop? Friction for the pavement is f=0.34 and he is traveling downhill at 4%.

a) 1005ft
b) 1120ft
c) 1450ft
d) 1675ft

Problem 14

An engineer is researching runoff from a farmer's field upstream of his new project site. A geotechnical report shows the area consists mainly of a silty loam with coarse to moderately fine textures. The farmer's field consists of 75% well-maintained (good hydraulic condition), straight row alfalfa and the rest has been converted to a local park with 80% grass cover. What is the initial abstraction of the farmer's land?

a) 0.71 inches
b) 0.74 inches
c) 1.70 inches
d) 2.08 inches

Problem 15

A freeway curve is having its speed limit increased from 50mph to 65mph. The existing freeway curves radius is 1750ft. What superelevation is required to accommodate the change in the speed limit?

a) .05
b) .06
c) .09
d) .10

Problem 16

A traffic study of a major arterial street was conducted during morning rush-hour traffic. Using the data from the study, what is the peak hour factor?

a) 0.76
b) 0.81
c) 0.84
d) 0.93

Time Interval	Vehicle Count
6:30 - 6:45	800
6:45 - 7:00	835
7:00 - 7:15	900
7:15 - 7:30	965
7:30 - 7:45	1025
7:45 - 8:00	925
8:00 - 8:15	750
8:15 - 8:30	715
8:30 - 8:45	625
8:45 - 9:00	585

Problem 17

An 8ft x 8ft footing carries a load of 1500 psf. What force is felt by the underlying soil 8ft away at a depth of 24 ft?

a) 60 psf
b) 150 psf
c) 500 psf
d) 1500 psf

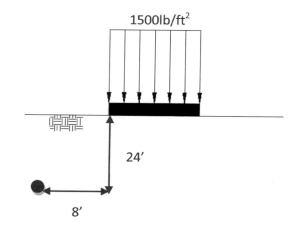

Problem 18

The sag vertical curve in the figure below is approaching an overpass. What is the clearance between the roadway and the overpass?

a) 16.8ft
b) 22.6ft
c) 27.2ft
d) 33.5ft

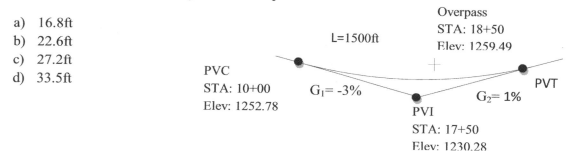

Problem 19

A vehicle is traveling on a highway with a design speed of 55mph. According to AASHTO, what is the stopping sight distance an engineer should use for their design?

a) 425 ft
b) 492.4 ft
c) 495 ft
d) 570 ft

Problem 20

Which of the following statements about roadway 'Levels of Service' is incorrect?

a) Level A represents unimpeded flow.
b) Economic considerations favor high traffic volumes and more obstructed levels of service.
c) A roadway's level of service can change throughout the day.
d) There are five different levels of service.

Problem 21

A track-hoe can excavate $5ft^3$ of ground per bucket. It takes the track-hoe 40 seconds to dig and place the load into a dump truck before it can start digging again. When the soil is dropped into the dump truck, it expands 15%. The dump trucks can hold a volume of $20yd^3$ of material. If the track-hoe operator takes two-15 minute breaks, how many dump trucks are needed to remove all material excavated in an 8-hour shift?

a) 6
b) 7
c) 8
d) 9

Problem 22

Which of the following standardized soil testing procedures is best for quickly determining field compaction results?

a) Standard Penetration Test
b) Modified Proctor Test
c) In-place Density Tests
d) Cone Penetrometer Test

Problem 23

A standard penetration test was used to help determine the soils at a project site. The image here portrays the in-situ soils conditions. What is the effective stress at point A?

a) 2100 psf
b) 1824 psf
c) 2475 psf
d) 1539 psf

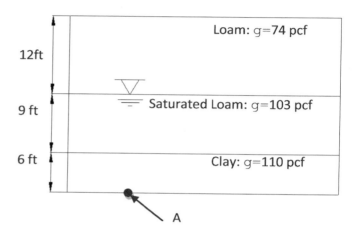

Problem 24

A contractor is purchasing a new track-hoe for $100,000 today. The track-hoe will effectively earn the contractor $40,000 year 1, $32,000 year 2, and $25,000 year 3. If the track-hoe is sold at the end of year 3 for $30,000, what is the loss/gain to the contractor presently (Today's value)? Assume an interest rate of 4%.

a) ($3,547)
b) $12,903
c) $16,942
d) $27,000

Problem 25

A custom home is being built upon a raised pad. Using the fill heights shown in the figure, how much fill is required? Assume the fill does not require side slopes.

a) 75 yd^3
b) 86 yd^3
c) 111 yd^3
d) 214 yd^3

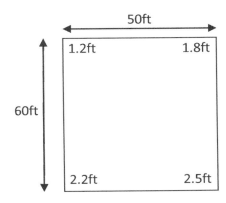

Problem 26

A cantilever beam experiences a force acting as an end moment shown in the figure. What is the maximum deflection in the beam?

E=29,000ksi

I=1650in^4

a) -2.50x10^4 in
b) -1.50x10^2 in
c) -0.036 in
d) 0 in

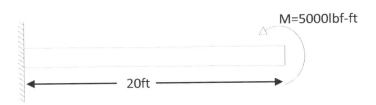

Problem 27

A rectangular beam has a 4in x 6in cross section. If the beam experiences a bending moment of 0.7 k-ft, what is the bending stress in the beam?

    a) 0 ksi
    b) 0.12 ksi
    c) 0 .25 ksi
    d) 0 .35 ksi

Problem 28

Which of the following statements about concrete is true?

    a) The hydraulic load on concrete formwork is greatest after the concrete sets.
    b) If the water-to-cement ratio of concrete is decreased, water tightness is decreased
    c) Adding water to concrete mix increases workability, increases slump, and can decrease the concretes strength.
    d) A pound of concrete weighs more than a pound of feathers.

Problem 29

The system in the figure to the right contains an incompressible fluid. What is the flow rate (Q) of pipe 3?

    a) 0.09 ft^3/s
    b) 0.19 ft^3/s
    c) 2.65 ft^3/s
    d) 8.9 ft^3/s

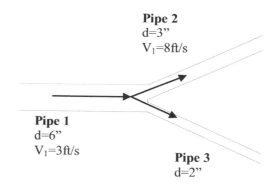

Pipe 2
d=3"
V_1=8ft/s

Pipe 1
d=6"
V_1=3ft/s

Pipe 3
d=2"

MIKE HANSEN, P.E.

Problem 30

An 8" PVC water system delivers water from a tank (elevation=1890ft) to a subdivision 3000 ft away. If the line is flowing at 2200gpm, what is the head loss from the tank to the subdivision? Assume C=150 for PVC pipe.

a) 25 ft
b) 145 ft
c) 180 ft
d) 335 ft

Problem 31

Rain falls onto the two adjacent areas shown in the figure. Using the rational method, what is the peak flow of the storm water runoff? Assume the runoff flows from Area 1 through Area 2.

a) 9 CFS
b) 12 CFS
c) 22 CFS
d) 47 CFS

Storm Intensity

time (min)	I (in/hr)
15	1.2
30	2.7
50	3.2

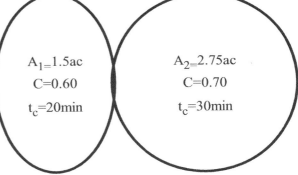

A_1=1.5ac
C=0.60
t_c=20min

A_2=2.75ac
C=0.70
t_c=30min

Problem 32

A concrete cylinder is being tested with a vertical loading. If the concrete breaks at a compressive stress of 4000psi, what was the vertical loading? The diameter of the concrete cylinder is 6 inches.

P=?

a) 113kips
b) 125kips
c) 250kips
d) 300kips

Problem 33

Poisson's ratio v is the ratio of lateral strain to the axial strain. When a sample object is stretched in one direction, it tends to contract somewhere else. Which of the following most aptly applies to Poisson's ratio?

a) Elastic Strain
b) Lateral Strain
c) Axial Strain
d) Tensile Strength

Problem 34

What is the Euler load for the slender vertical column with pinned ends?

$E = 2.9 \times 10^7$ lbf/in^2

$I = 0.5$ in^4

a) 9.9 kips
b) 19.8 kips
c) 54 kips
d) 117 kips

H = 10ft

Problem 35

A concrete beam is being designed for use with 2.0 in^2 of steel rebar. The concrete will be a 3000-psi mix (f'_c) and the tension steel yield strength (f_y) is 60,000 lbf/in^2. What area of concrete (Ac) is required to balance the steel?

a) 24in^2
b) 35in^2
c) 41in^2
d) 47in^2

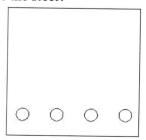

Problem 36

A gravity retaining wall is supporting saturated clay with the properties shown in the figure. What is the total active resultant per unit width of wall?

a) 17 kips/ft
b) 19 kips/ft
c) 20 kips/ft
d) 24 kips/ft

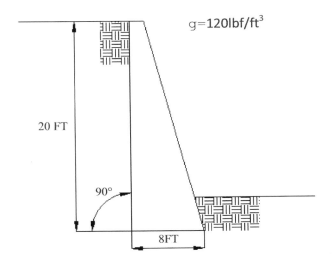

Problem 37

If the water in the figure were to dry up, which of the following statements is true about the cohesive factor of safety? $\gamma_{sat} = 110$ lb/ft^3, γ_{dry}=90 lb/ft^3

a) The cohesive factor of safety increases.
b) The cohesive factor of safety stays the same.
c) The cohesive factor of safety decreases.
d) There is not enough information to make a determination.

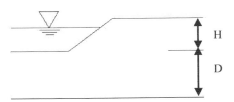

Problem 38

According to the concrete mix in the table below, what is the total volume of the mix?

a) 0.5 yd^3
b) 1.0 yd^3
c) 1.75 yd^3
d) 2.25 yd^3

	SSD Weight	Specific Gravity
Low Alkali Cement	442	3.15
Class F Fly Ash	94	2.1
Coarse Aggregate	1809	2.57
Fine Aggregate	1221	2.6
Potable Water	300	1
Total Air	1.50%	-

Problem 39

A soil sample with a volume of 1ft^3 was determined to be 50% saturated and 10% porous. If the specific gravity of the soil is 2.10, what is the dry unit weight of the soil?

a) 79 lb/ft^3
b) 102 lb/ft^3
c) 118 lb/ft^3
d) 167 lb/ft^3

Problem 40

A city with 50,000 people is designing a new wastewater plant to accommodate its current population. The average family is 2.5 persons. If the average amount of solids is 500mg/L, what should the plant be sized for daily intake?

a) 80 lbm/day
b) 2000 lbm/day
c) 26,000 lbm/day
d) 50,000 lbm/day

Example Problem Solutions

Problem 1 - Solution

A retaining wall is supporting a soil with the properties shown in the figure. What is the total active resultant per unit width of wall?

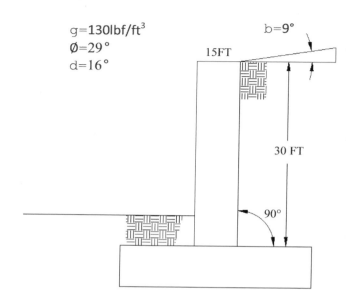

$g=130 lbf/ft^3$
$\emptyset=29°$
$d=16°$

a) 18 kips/ft
b) **21 kips/ft**
c) 24 kips/ft
d) 30 kips/ft

Step 1: Problem Analysis

- This problem is an active earth pressure problem. Geotechnical
- The problem asks to solve for the <u>total active resultant</u> per unit width of wall.

Step 2: Reference

- The equation for total active resultant is found on page 37-4, equation 37.10(b)
- The formula is $R_a=\frac{1}{2} k_a g H^2$. The only variable we are missing is k_a.
- We have two theories for active earth problems to solve for k_a, Rankine or Coulomb.
- Coulomb theory is used for problems involving friction (d), a sloping backfill (angle b), and an inclined active-side wall face (angle θ).
- Rankine theory disregards wall friction.
- k_a is solved using the Coulomb theory because we have d, b, and θ.
- We use the Coulomb theory because we have friction in our problem.
- The Coulomb formula is on page 37-3 in the CERM, equation 37.5.

Tab CERM

Tab CERM Pg 37-4
Label: Resultant Force

Tab CERM Pg 37-3
Label: Coulomb

Step 3: Solve for K_a, equation 37.5

- $$K_a = \frac{\sin^2(\theta+\emptyset)}{\sin^2\theta \sin(\theta-\delta)\left(1+\sqrt{\dfrac{\sin(\emptyset+\delta)\sin(\emptyset-\beta)}{\sin(\theta-\delta)\sin(\theta+\beta)}}\right)^2}$$

- $\theta = 90°$ because the wall is vertical.
- Given Data: $\emptyset= 29°$, $g =130\text{lbf/ft}^3$, d=16°,b=9°

- $$K_a = \frac{\sin^2(90+29)}{\sin^2(90) \sin(90-16)\left(1+\sqrt{\dfrac{\sin(29+16)\sin(29-9)}{\sin(90-16)\sin(90+9)}}\right)^2} = \underline{0.366}$$

Step 4: Solve for R_a

- $R_a = \frac{1}{2} k_a g\ H^2 = \frac{1}{2}(0.366)*(130\text{lbf/ft}^3)*(30\text{ft})^2 = 21{,}411$ lbf/ft or **b) 21.4 kips/ft**

Problem 2 - Solution

In asphalt paving, a thicker section of pavement (t >4") has many advantageous benefits. Which of the following is not one of the benefits?

a) Lifts can be placed in cooler weather
b) More economical to place one lift versus two
c) It can be easier to reach density requirements
d) **Requires less time to setup before drivers can use it**

Step 1: Problem Analysis

- Asphalt paving - Traffic
- Which of the answers is false?

Step 2: Reference

- Section 75 discusses Flexible Pavement Design
- More specifically, the benefits of a thicker pavement section are found on Pg 75-2.
- If you look in the index under pavement, asphalt pavement is on Pg 75-1, if you looked there, you would have found the answer on the following page, 75-2.
    - o The index is your best friend for theory problems.

Tab CERM

Tab CERM Pg 75-2
Label: Asphalt

Step 3: Solve for the false answer

a) Lifts can be placed in cooler weather – True, Found on Pg 75-2
b) More economical to place one lift versus two - True, Found on Pg 75-2
c) It can be easier to reach density requirements - True, Found on Pg 75-2
d) **Thicker layer requires less time to setup before drivers can use it – False. A thicker layer with more heat would take longer to cool down and setup.**

Note: This type of problem is considered a 'Theory' question. Theory questions if not answered quickly can easily cause you a lot of frustration. Utilize the index.
1. Look up the topic, "pavement"
2. Asphalt pavement is on page 75-1, check there.
3. If you cannot find what you are looking for within a minute, move on and go back to the problem later.

Problem 3 - Solution

A 6ft x 8ft reinforced concrete box culvert (n=0.013) receives a flow of 215cfs from an irrigation canal. The slope of the culvert is 2%. Assume the culvert does not flow full. What is the flow depth in the box culvert?

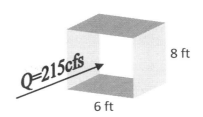

8 ft

6 ft

a) **2ft**
b) 3ft
c) 5ft
d) 6ft

Step 1: Problem Analysis

- This problem involves open channel flow. It is open channel flow because the problem states that the box culvert is not full. - Hydrology
- Find the flow depth, d

Tab CERM

Tab CERM Pg 19-4
Label: Manning's Eq.

Step 2: Reference

- Open channel flow begins on pg 19-3.
- Manning's equation is the go-to equation for open channel flow.
- Pg 19-4, Eq. 19.13(b)

Step 3: Break down A and R into terms of depth, d

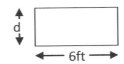
d

6ft

- Manning's equation is $Q = \frac{1.49}{n} A R^{2/3} \sqrt{S}$
- In our problem, $A = 6 \times d$, The A in Manning's equation is the <u>flow area</u>, not the area of the box culvert.
- We also need R in terms of d. R is the hydraulic radius and a table on Pg 19-3 gives hydraulic radius for common shapes.
- $R = \frac{wd}{w+2d} = \frac{6d}{6+2d}$

Step 4: Solve for depth, d

- $Q = \frac{1.49}{n} A R^{2/3} \sqrt{S}$

- $Q = \frac{1.49}{n} (6 \times d)(\frac{6d}{6+2d})^{2/3} \sqrt{S}$ Substitute A and R in terms of d

- $215 = \frac{1.49}{0.013} 6d(\frac{6d}{6+2d})^{2/3} \sqrt{0.02}$

- $2.21 = d(\frac{6d}{6+2d})^{2/3}$

- Unfortunately, without a graphing calculator to solve this, you must use trial and error. Start with one of the values offered in the multiple-choice answers. Generally, B is a good first choice. If the value is lower, then the answer is probably (a). If it is higher than you have one more calculation to determine if it is C or D. It is always a good idea to check that your answer makes the equation work.

- Let us try b) 3ft.

- $2.21 = 3(\frac{6*3}{6+2*3})^{2/3} = 3.93$. This does not work, 2.21 does not equal 3.93.

- Try a) 2ft.

- $2.21 = 2(\frac{6*2}{6+2*2})^{2/3} = 2.258$. These values are very close. **a) 2ft is the answer.**

Problem 4 - Solution

The pump system in the figure is drawing water from the lake and pumping it into the tank. The pump efficiency is 90%, the Darcy friction factor is 0.02, the diameter of the pipe is 4 inches, pipe length is 100 ft, and flow rate is 240 Gallons per minute.

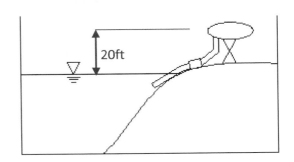

a) 0.5 hp
b) 1.0 hp
c) 1.5 hp
d) 2.5 hp

Step 1: Problem Analysis

- This problem involves an energy system, energy loss due to friction, and kinetic energy. Hydrology
- Find the Horsepower of the pump

Step 2: Reference

- Bernoulli's equation is used for energy conservation problems. The easiest way to attack the problem is to think of the entire system. The tank head and the energy loss due to friction (velocity through pipe) are working against the pump.

  Pump Head = Tank head + Energy Loss due to Friction

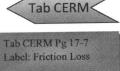

- Pg. 17-7, Eq 17.28 covers friction loss.
  - $hf = \dfrac{fL}{D}$
- Pg. 16-2, Eq. 16.3(b) is Kinetic Energy
  - $Ev = \dfrac{v^2}{2g}$
- To add friction to the kinetic energy the two equations are multiplied together.
  - Friction loss and flow through the pipe $= \dfrac{fLv^2}{2Dg}$

- Pg. 18-8, Table 18.5 contains Hydraulic horsepower Equations
  - A modification of the equation and incorporating efficiency yields; $\quad HP = \dfrac{\gamma Q H}{550e}$

Tab CERM

Tab CERM Pg 18-8
Label: Horsepower

Write this HP formula in
your CERM.

Step 3: Solve for Pump Head

- Pump Head $= H_{tank} + \dfrac{fLv^2}{2Dg} = 20 + \dfrac{0.02(100)v^2}{2\left(\frac{4}{12}\right)32.2}$

- We need velocity, $v = \dfrac{Q}{A} = \dfrac{240 gal/min}{\pi\left(\frac{4}{12}\right)^2/4} \times \dfrac{0.133601 ft3}{1\ gal} \times \dfrac{1\ min}{60\ sec} =$

  6.12ft/s

- Pump Head $= 20 + \dfrac{0.02(100)(6.12)^2}{2\left(\frac{4}{12}\right)32.2} = 23.49\text{ft}$

Step 4: Solve for Horsepower

- $HP = \dfrac{\gamma Q H}{550e} = \dfrac{62.4(0.53)(23.49)}{550(0.90)} = \quad$ **c) 1.56 hp**

MIKE HANSEN, P.E.

Problem 5 - Solution

A new roadway is to be constructed and the contractor is preparing grade. According to the stations and cut/fill shown in the table, what is the amount of excavation required?

Station	Fill(+) or Cut (-)
1+00	$+50ft^2$
1+50	$-15ft^2$
2+50	$-75ft^2$
4+00	$0ft^2$

a) $225yd^3$
b) 343 yd^3
c) 650 yd^3
d) 1025 yd^3

Step 1: Problem Analysis

- This is a construction earthwork problem.
- Calculate the amount of excavation required for final grading.

Step 2: Reference

- To solve this problem we use the average end area method (AEA).
  - You use the AEA method when you have a fill/cut and each station is given. Unless the problem asks specifically for you to use another method, this is the quickest and easiest way to solve the problem.
- Pg. 79-4, Eq. 79.58

  - $$V = \frac{L(A1+A2)}{2}$$

Tab CERM

Tab CERM Pg 79-4
Label: Avg End Area

Step 3: Solve for the excavation between stations

- Between 1+00 – 1+50: $\dfrac{50ft(50ft^2 +(-15ft^2))}{2} = 875ft^3$
- Between 1+50 – 2+50: $\dfrac{100ft(-15ft^2+(-75ft^2))}{2} = -4500ft^3$
- Between 2+50 – 4+00: $\dfrac{150ft(-75ft^2+0)}{2} = -5625ft^3$
- Total = -9250 ft^3
- The problem asks for yd^2, so convert your solution.
  - -9250 ft^2 x 1 yd^3 / 27 ft^3 = **b) 343 yd^3**

The Burm Question: What if they asked you to use the Prismoidal Formula Method to solve the problem?

- The PFM is found on Pg 79-4. A_m is the average area in-between two stations.

Problem 6 - Solution

Which of the following statements about
open channel cross sections is true?

a) An efficient open channel cross section minimizes flow.
b) The most efficient rectangle cross section has a depth equal to one third the width: $d=\frac{w}{3}$
c) A semicircular cross section is the most efficient shape and minimizes construction cost.
d) **The most efficient cross section requires the hydraulic radius be maximum and the wetted perimeter minimized.**

Step 1: Problem Analysis

- We have a theory question involving the efficiency of an open channel cross section.
- We need to determine which of the statements [A-D] are true.
  There will be only one true answer.

Step 2: Reference

- In this theory question, you are lucky, because the answer comes right out of the CERM. Some theory questions may require you to just know the answer or be able to deduce what the solution should be.
- Take your time to verify or disprove each answer, remember you still have six minutes per problem.
- Pg. 19-9, section 12: Most Efficient Cross Section.

Tab CERM Pg 19-9 Label:
Most Efficient cross-section

Step 3: Find the correct answer

- Read each of the statements and prove them incorrect or correct from the information available in the CERM.

a) An efficient open channel cross section minimizes flow. – Incorrect, It maximizes flow.
b) The most efficient rectangle cross section has a depth equal to one third the width: $d=\frac{w}{3}$: Incorrect. $d=\frac{w}{2}$
c) A semicircular cross section is the most efficient shape and minimizes construction cost. – Incorrect ; A semicircular cross section is the most efficient cross section, however it is generally the hardest and most expensive to construct.
d) **This answer is true.**

Problem 7 - Solution

A sieve analysis was completed on native soil where a homebuilder is proposing a new subdivision. What is the USCS classification for this sample?

Sieve Number	Percent Finer
1.5 inch	98
No. 4	85
No. 10	72
No. 20	68
No. 40	62
No. 100	57
No. 200	52

a) SW
b) CH
c) MH
d) CL

Liquid Limit = 55

Plastic Limit = 27

Step 1: Problem Analysis

- Unified Soil Classification System
- We are solving for the USCS soil classification

Step 2: Reference
- THE USCS table and chart is on Pg. 35-6, table 35.5

Tab CERM

Tab CERM Pg 35-6
Label: USCS

Step 3: Solve for the soil classification

- What percent is passing the #200 sieve? We have over 50% passing, according to the USCS table it is a fine-grained soil.
- The liquid limit (LL) is 55, so we have a high compressibility. This leaves us with MH, CH, or OH. Two of those answers are in the multiple-choice answers, keep breaking it down.
- Now we need to use the chart to determine our soil. We need to calculate the plasticity index (PI).
- Pg. 35-5, shows PI $= LL - PL$
    - LL = 55 (given), PL = 27 (given)
    - PI = 55-27 = 28
  - Since we have the PI and LL, we can look at the chart to determine the soil type. It appears we have a CH, but we are close to the A-line. Calculate the A-line to verify we are above it.
  - A-line = 0.73(LL-20) = 0.73(55-20) = 25.55, since our PI is 28 we know we are above the A-line and we do in fact have a **<u>CH.</u>**

Problem 8 - Solution

Which of the shear diagrams most-accurately resembles the beam loading shown in the figure?

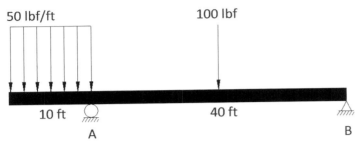

Step 1: Problem Analysis

- Determinate Statics - Structures

- Which Shear Diagram is correct?

Step 2: Reference

- Shear and Bending Moment Diagrams are on Pg. 44-8, section 12.
- First step is to calculate the reactions at the supports A and B. We will be using the conditions of equilibrium from Pg. 41-6, section 17.
- Using the reactions, we will then draw the diagram.

Tab CERM

Tab CERM Pg 44-8
Label: Shear and
Bending Moment Diag.

Step 3: Solve for reactions at the supports.

- $\sum MB = 0 = 500\text{lbf} \times 45\text{ft} - A \times 40\text{ft} + 100\text{lbf} \times 20\text{ft}$
  - A=612.5lbf
- $\sum Fy = 0 = -500\text{lbf} + A - 100\text{lbf} + B$
  - $= -500\text{lbf} + 612.5\text{lbf} - 100\text{lbf} + B$
  - B = -12.5lbf, the reaction at B is downward.

The distributed load to the left of support A is calculated by applying a single load of 500lb (50*10) at the midpoint.

Step 4: Draw the Shear Diagram

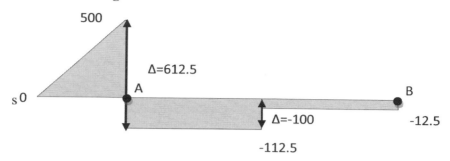

- Start drawing the shear diagram from support B. We know the reaction at B is -12.5lbf. Once we reach the single 100lb load, we add that negative force to our shear to get -112.5lbf. Once we get to support A, we add the positive reaction of 612.5lbf. To the left of support A, we subtract 50lbf every foot giving us the shear triangle that goes to zero at the free end.
- **Answer is C**

The Burm Question: What if they ask you to determine the bending moment diagram of the beam loading?

- Pg. 44-8 has an example problem that shows how to calculate and draw a bending moment diagram.

Problem 9 - Solution

A contractor is looking to build a headwall for a 36" storm drain outlet. How much plywood should he purchase for forming the headwall structure? Assume there is 10% waste over the exact calculation. The headwall is 1ft thick.

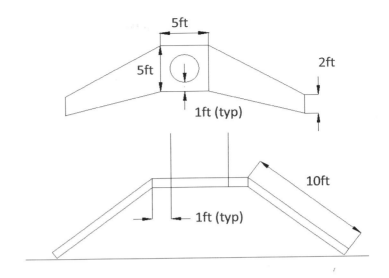

a) 89.93 ft^2
b) 98.93ft^2
c) 179.86 ft^2
d) **193.45 ft^2**

Step 1: Problem Analysis

- Construction Formwork
- Calculate the amount of plywood including waste that needs to be ordered to form the structure.

Step 2: Reference

- There is no need to reference the CERM. This is a test on your mathematics skill.

Step 3: Solve for the areas that require forms.

- 5ft x 5ft = 25ft^2. We have to form the front and back, 25 ft^2 x 2 = 50ft^2

- Subtract for the pipe.

  o $\dfrac{\pi D^2}{4} = \dfrac{\pi 3^2}{4} = 7.07$, front and back = 7.07 x 2 = (- 14.14 ft^2)

- Add in the wing walls.
  - ○ 10ft x 2ft + (3ft x 10ft)/2 = 35ft^2. We have the front and back of two walls. 35ft^2 x 4 = 140ft^2

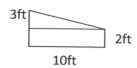

- Add everything up: 50 ft^2 + (-14.14 ft^2) + 140 ft^2 = 175.86ft^2
- Adjust for waste: 179.86 ft^2 x 1.10 = **d) <u>193.45 ft^2</u>**

Problem 10 – Solution

Use the construction schedule in the table to determine the duration of the critical path.

a) 13 days
b) 15 days
c) 17 days
d) 18 days

Activity	Duration (days)	Predecessor
A - Excavate Box Culvert	4	-
B - Excavate Turn Downs	1	A
C - Place Footing Forms	2	A
D - Place Footing Rebar	3	A
E - Pour Footing	1	B,C,D
F - Place Box Culvert Rebar	5	E
G - Set Box Culvert Forms	3	E
H - Pour Box Culvert	1	F,G
I - Backfill Structure	3	H

Step 1: Problem Analysis

- Construction Scheduling
- Find the duration of the critical path

Step 2: Reference
- Pg. 85-11 has a good example on construction scheduling and also describes early start, early finish, late start, and late finish. These may be asked for on the PE Exam.
- In this problem, you need to draw up the schedule detailed in the table and determine the longest project duration.

Tab CERM

Tab CERM Pg 85-11
Label: Scheduling

Step 3: Sketch up the schedule, durations, and predecessors for each of the activities

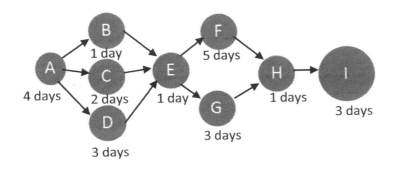

Step 4: Solve for the critical path duration

- The critical path in this schedule is ADEFHI. Each of the activities has to be completed within the duration or the project is extended. Take the duration of each of these activities and add them up.
  - 4+3+1+5+1+3 =**c) 17 days.**

The Burm Question:

- What if the exam asks the following; How much float is there along path ABEGHI?

Solution: Float is the amount of days the activity can not be worked and still have the project finished at the same time. In this case path ABEGHI = 4+1+1+3+1+3 = 13 days. So there would be 4 days of float (17-13=4).

*Study Note: Be familiar with calculating early start, late start, early finish, and late finish. This may be on the exam. For example; activity B: Early start Day 4. Early Finish is day 5. If we use the available float (2 days from task D), then the Late start is Day 6 and the Late finish is Day 7.

Problem 11 - Solution

The force in member BC is most nearly?

a) 11k Compression
b) **5k Compression**
c) 1k Tension
d) 11k Tension

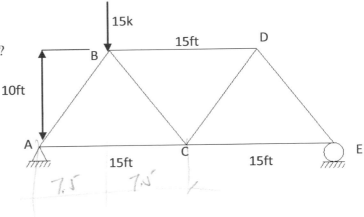

Step 1: Problem Analysis

- We have a determinate truss
- We are solving for the force in member BC

Step 2: Reference

- Trusses are discussed within Pg.41-13 to 41-16.
- We use the cut and sum method on this problem to determine the force in member BC.
- We first solve for the reactions at the supports and then make a vertical cut through member BC. The sum of all forces in the Y direction = 0

Tab CERM Pg 41-13
Label: Trusses

Step 3: Solve for reactions at the supports

- $\sum M_E = 0 = -A_y(30\text{ft}) + 15k(22.5\text{ft})$
  - o $A_y = 11.25k$

Step 4: Solve for the force in member BC

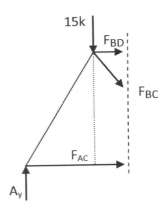

- Draw a free body diagram of the truss and make a vertical cut through member BC.
- $\theta = \tan^{-1}(\frac{7.5}{10}) = 36.87°$
- $\sum F_y = 0 = A_y - 15k - F_{BC}*\cos(36.87)$
- $F_{BC} = -4.69k$ **(Compression)**, It is compression because the value is negative. If it were positive, it would be in tension.

Problem 12 - Solution

A horizontal curve with a radius of
1500ft has the PI shown in the figure.
What is the station of the PT?
I=14°

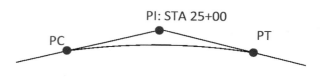

PI: STA 25+00

PC

PT

a) 13+41
b) 14+67
c) 25+33
d) 26+83

Step 1: Problem Analysis

- Horizontal Curve - Traffic
- Calculate the station of the PT

Step 2: Reference

- Horizontal Curves are on pages 78-1 to 78-7.
- There are many useful equations for solving horizontal curve problems throughout these pages. I recommend putting together a sheet that has all of these equations on them. Example: $R = \dfrac{5729.578}{D}$, $L = \dfrac{2\pi RI}{360°}$

Tab CERM

Tab CERM Pg 78-1
Label: Horiz. Curves

Step 3: Solve for the Station of the PT (STA PT)

- Our problem asks us to solve for the station of the PT. The first step is to look into what equation gives us the PT.
  - Eq. 78.11, pg 78-3, sta PT= sta PC+ L
- So we need to know the sta PC and L.
  - Eq. 78.12, pg 78-3, sta PC = sta PI – T
- We know the sta PI and can calculate T
  - Eq. 78.4, pg 78-2, $T = R \tan\dfrac{I}{2}$
  - T = 1500ft * tan(14°/2) = 184.18ft
  - PC= (25+00) – 184.18ft = 2500-184.18 = 2315.82 or 23+16
- Now just calculate L to solve the problem.
  - Eq. 78.3, pg 78-2, $L = \dfrac{2\pi RI}{360°} = \dfrac{2\pi 1500(14)}{360°} = 366.52$
  - STA PT = STA PC+L = (23+16)+366.52 = **d) STA 26+83**

Problem 13 - Solution

A sports car is speeding down a straightaway at 80MPH when he notices a stop sign rapidly approaching. Considering his perception reaction time (assume 2.5 seconds), how far will he travel before the car comes to a stop? Friction for the pavement is f=0.34 and he is traveling downhill at 4%.

   a) **1005ft**
   b) 1120ft
   c) 1450ft
   d) 1675ft

Step 1: Problem Analysis

   - Sight Stopping Distance - Traffic
   - Calculate the stopping distance including perception reaction time

Step 2: Reference

   - Pg. 78-9, section 13 covers Stopping Sight Distance
      o Eq. 78.43(b), $S=1.47t_{sec}*v_{mph} + \dfrac{v^2{}_{mph}}{30(f+G)}$

Tab CERM

Tab CERM Pg 78-9
Label: SSD

Step 3: Solve for SSD

   - $S=1.47t_{sec}*v_{mph} + \dfrac{v^2{}_{mph}}{30(f+G)} = 1.47(2.5sec)*(80mph) + \dfrac{80mph^2}{30(0.34-.04)} =$ **a) 1005.1ft**
   - G= - .04 because it is downhill.

Problem 14 - Solution

An engineer is researching runoff from a farmer's field upstream of his new project site. A geotechnical report shows the area consists mainly of a silty loam with coarse to moderately fine textures. The farmer's field consists of 75% well-maintained (good hydraulic condition), straight row alfalfa and the rest has been converted to a local park with 80% grass cover. What is the initial abstraction of the farmer's land?

    a) 0.71 inches
    b) 0.74 inches
    c) 1.70 inches
    d) 2.08 inches

Step 1: Problem Analysis

- This is a water resources problem involving NRCS curve numbers.
- Calculate the initial abstraction of the area

Step 2: Reference

- Relevant pages in the CERM are 20-15 to 20-19.
- Pg 20-16 has a step-by-step process how to handle an advanced NRCS problem, however you will generally only need the first step for the PE exam.
- Use step 1 to determine your soil classification of the area.
- Then use tables 20.4 and 20.5 to determine the CN numbers for the land your problem includes.

Tab CERM

Tab CERM Pg 20-16
Label: Initial Abstraction

Step 3: Solve for Soil Classification

- The problem states we have a silty loam with coarse to moderately fine textures. Reading through the A through D classifications(Step 1) this is clearly a type B soil.

Step 4: Solve for CN values

- We have two different types of land in this problem.
    o Straight Row Crop – good hydrologic condition, Table 20.5 **(Alfalfa)**
    o Open space – 75%+ grass cover, Table 20.4 **(Park)**
    o Note the two different tables.

	Soil Classification	CN	Percent of Area	Weighted CN
Alfalfa	B	78	0.75	58.5
Park	B	61	0.25	15.25
				73.75

Step 5: Solve for Initial Abstraction

- The initial abstraction corresponds the CN number and is found in table 20.6.
- With a curve number (CN) of 73.75 we take the weighted average of 73 and 74.

CN	Initial Abstraction	Weight	Weighted Initial Abstraction
73	0.74	0.25	0.19in
74	0.703	0.75	0.53in
			0.71in

- **The initial abstraction for the farmers land is a) 0.71 inches.**

Problem 15 - Solution

A freeway curve is having its speed limit increased from 50mph to 65mph. The existing freeway curves radius is 1750ft. What superelevation is required to accommodate the change in the speed limit?

    a) .05
    b) .06
    c) .09
    d) .10

Step 1: Problem Analysis

- Transportation - Superelevation
- Calculate the required superelevation

Step 2: Reference

- Pg. 78-6, section 10 covers superelevation
- Equation 78.37(b) will be the most common equation you will need to use on the PE Exam.

> Tab CERM

> Tab CERM Pg. 78-6
> Label: Superelevation

- $e = \dfrac{v_{mph}^2}{15R} - f_s$, This equation uses velocity in units of mph. f_s= side friction factor.

- $50mph \leq velocity \leq 70mph$, use $f_s = 0.14 - \dfrac{0.02(v_{mph}-50)}{10}$ [eq. 78.39]

- $30mph < velocity < 50mph$, $f_s = 0.16 - \dfrac{0.01(v_{mph}-30)}{10}$ [eq. 78.38]

- $velocity \leq 30mph$, use f_s=0.16

Step 3: Solve for side friction factor

- The velocity is 65 mph, so we use equation 78.39
- $f_s = 0.14 - \dfrac{0.02(65-50)}{10} = 0.11$

Step 4: Solve for superelevation

- $e = \dfrac{65^2}{15(1750)} - 0.11 = 0.05$
- e =0.05

MIKE HANSEN, P.E.

Problem 16 - Solution

A traffic study of a major arterial street was conducted during morning rush-hour traffic. Using the data from the study, what is the peak hour factor?

a) 0.76
b) 0.81
c) 0.84
d) 0.93

Time Interval	Vehicle Count
6:30 - 6:45	800
6:45 - 7:00	835
7:00 - 7:15	900
7:15 - 7:30	965
7:30 - 7:45	1025
7:45 - 8:00	925
8:00 - 8:15	750
8:15 - 8:30	715
8:30 - 8:45	625
8:45 - 9:00	585

Step 1: Problem Analysis

- Traffic – Peak hour factor
- Calculate the Peak Hour Factor

Step 2: Reference

- How to calculate the PHF is found on Pg. 73-5
- To calculate the PHF, we need determine the peak hour volume, and the peak 15 minute volume.

  o $$PHF = \frac{V_{vph}}{4V_{15\,min,peak}}, \text{ eq. 73.5}$$

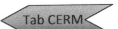

Tab CERM Pg 73-5
Label: Peak Hour Factor

Step 3: Solve for peak hour volume

Time Saver Tip: Go down the list and find the 4 consecutive highest numbers. You can see 7-8am has the largest values. If it is too difficult to determine it by looking at the numbers, then calculate them all out.

Hour Interval	Add 15 min intervals	Peak Hour Volume
6:30-7:30	800+835+900+965	3490
6:45-7:45	835+900+965+1025	3725
7:00-8:00	900+965+1025+925	**3815**
7:15-8:15	965+1025+925+750	3665
7:30-8:30	1025+925+750+715	3415
7:45-8:45	925+750+715+625	3015
8:00-9:00	750+715+625+585	2675

- The peak hour volume is 3815 vehicles

Step 4: Solve for the 15-minute peak volume

- Simply look at the 15-minute intervals and take the largest value. In this case it is the interval 7:30-7:45, 1025 vehicles.

Step 5: Solve for Peak Hour Volume

- $PHF = \dfrac{3815}{4*1025} = 0.93$

-

- **PHF = d)** <u>**0.93**</u>

Problem 17 - Solution

An 8ft x 8ft footing carries a load of 1500 psf. What force is felt by the underlying soil 8ft away at a depth of 24 ft?

a) **60 psf**
b) 150 psf
c) 500 psf
d) 1500 psf

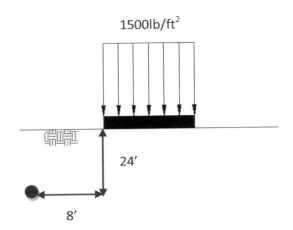

Step 1: Problem Analysis

- Geotechnical - Boussinesq

- Calculate the force felt by underlying soil

Step 2: Reference
- Boussinesq's equation is used to determine the increase in vertical stress caused by a load on the soil surface.
- The easiest way to solve this problem is using the Boussinesq chart in appendix A-71.

Step 3: Solve for the force applied to underlying soil

- Use the Boussinesq chart for square foundation (8x8 footing).
- A distance 8ft away and at 24ft deep (as described in problem) results in an increased force in the soil of 0.04p. The chart describes the width of the footing as a distance b. So we look 1b to the side and 3b downward. (3b=24ft)
- Simply take the force per square foot applied and multiply it by 0.04.
  - $1500 \text{lb/ft}^2 \times 0.04 = 60 \text{lb/ft}^2$
- The increase felt by underlying soil is **60 psf**.

  If the footing was 8x20, then you use the chart on the left for "infinitely long foundation." I know it is not really an infinite footing, but that is how the chart works.

The Burm Question: What is the footing was 8' x 100'?

Hint: Use the continuous footing table.

Problem 18 - Solution

The sag vertical curve in the figure below is approaching an overpass. What is
the clearance between the roadway and the overpass?

a) 16.8ft
b) 22.6ft
c) 27.2ft
d) 33.5ft

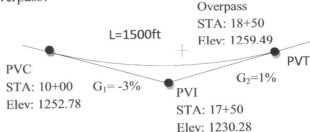

Step 1: Problem Analysis

- Vertical Curve
- Calculate the clearance or differences in elevations between the
  roadway and the overpass

Step 2: Reference

- The CERM references vertical curves on Pg. 78-10
- We use equation 78.47 to find the elevation of a point on a vertical curve.
  The x in the formula is in stations. If you are 750ft away from the PVC, then
  x=7.5.

  Tab CERM

  Tab CERM Pg 78-10
  Label: Vertical Curves

  o $Elev_x = \frac{R}{2}x^2 + G_1 x + elev_{PVC}$ (78.47)

- To calculate the radius, use equation 78.46

  o $R = \frac{G_2 - G_1}{L}$

- G1 and G2 are the grades as whole numbers (2% = 2) shown in the
  figure

Step 3: Solve for R and the Elevation at the overpass

$$R = \frac{G_2 - G_1}{L} = \frac{1-(-3)}{1500} = 0.266$$

- $Elev_x = \frac{0.266}{2}8.5^2 + (-3)8.5 + 1252.78 = 1236.89ft$

Step 4: Solve for the clearance

- $Elev_{overpass} - Elev_{roadway} = Clearance$
- 1259.49ft − 1236.89ft = **b) 22.6ft**

Problem 19 - Solution

A vehicle is traveling on a highway with a design speed of 55mph. According to AASHTO, what is the stopping sight distance an engineer should use for their design?

- a) 425 ft
- b) 492.4 ft
- **c) 495 ft**
- d) 570 ft

Step 1: Problem Analysis

- AASHTO – Stopping Sight Distance
- Determine the design SSD

Step 2: Reference

- Stopping sight distance can be calculated, however all this problem requires you to do is use table 78.2. This table is set up for typical design speeds and gives calculated and designed stopping sight distances based on AASHTO assumptions.

Tab CERM

Tab CERM Pg. 78-10
Label: AASHTO SSD

Step 3: Use the table to solve for SSD

- Use the table for 55mph. The design distance in feet is **c) 495**.

Problem 20 - Solution

Which of the following statements about roadway Levels of Service is incorrect?

a) Level A represents unimpeded flow.
b) Economic considerations favor high traffic volumes and more obstructed levels of service.
c) A roadway's level of service can change throughout the day.
d) **There are five different levels of service.**

Step 1: Problem Analysis

-   Levels of Service
-   Which of the statements is false?

Step 2: Reference
-   The CERM discusses levels of service on pg. 73-3

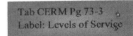

Tab CERM

Tab CERM Pg 73-3
Label: Levels of Service

Step 3: Solve for the false answer

-   Level A represents unimpeded flow. - **True**
-   Economic considerations favor high traffic volumes and more obstructed levels of service. - **True**
-   A roadway's level of service can change throughout the day. - **True**
-   There are <u>five</u> different levels of service. – **False, there are 6 levels of service (A-F)**

Problem 21 - Solution

A track-hoe can excavate $5ft^3$ of ground per bucket. It takes the track-hoe 40 seconds to dig and place the load into a dump truck before it can start digging again. When the soil is dropped into the dump truck, it expands 15%. The dump trucks can hold a volume of $20yd^3$ of material. If the track-hoe operator takes two-15 minute breaks, how many dump trucks are needed to remove all material excavated in an 8-hour shift?

    a) 6
    b) 7
    c) 8
    d) 9

Step 1: Problem Analysis

- Construction Volumes
- Calculate the amount of trucks to deliver the excavated material

Step 2: Reference

- There is no need to reference this problem to the CERM. It is a mathematics problem.

Step 3: Solve for the volume of material moved by the excavator

- The excavator removes $5ft^3$ every 40 seconds.
- If the track-hoe operator works an 8-hour shift with 2-15 minute breaks the effective work day is 7.5 hours.

- $$\frac{5ft^3}{40\ seconds} = \frac{xft^3}{7.5\ Hours\ x60\frac{min}{hr}\ x\ 60\frac{sec}{min}}, \text{x=3375ft}^3$$

- $3375ft^3 / 27 = 125yd^3$
- $125yd^3$ x 1.15(expansion) = $143.75yd^3$

Step 4: Solve for trucks required to remove material

- $143.75yd^3 / 20yd^3$ (per truck) = 7.18 or **8 trucks** total to remove all the material.

Problem 22 - Solution

Which of the following standardized soil testing procedures is best for quickly determining field compaction results?

a) Standard Penetration Test
b) Modified Proctor Test
c) In-place Density Tests
d) Cone Penetrometer Test

Step 1: Problem Analysis

- Geotechnical – Soils testing
- Which of the tests are best for quickly determining field compaction results?

Step 2: Reference
- Common soils testing procedures begin on page 35-17.

Tab CERM

Tab CERM Pg 35-17
Label: Soils Testing

Step 3: Solve for the best method

a) Standard Penetration Test – Requires a large machine and operator to determine what type of soil is in the ground. – Not the answer
b) Modified Proctor Test – This test measures compaction but required a sample from the field be brought in to be tested with a 10lbm hammer that falls 18 inches. – Not the answer
c) In-place Density Tests – Typically known as the field density test, a common compaction testing method in the field is using a nuclear gauge. – Correct Answer
d) Cone Penetrometer Test – The cone penetration test is used to classify soils. – Not the answer

Problem 23 - Solution

A standard penetration test was used to help determine the soils at a project site. The image here portrays the in-situ soils conditions. What is the effective stress at point A?

a) 2100 psf
b) 1824 psf
c) 2475 psf
d) **1539 psf**

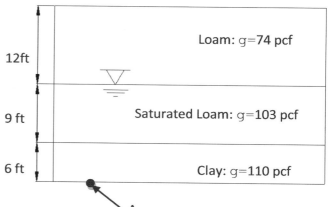

Step 1: Problem Analysis

- Geotechnical – Effective Stress
- Calculate the effective stress at point A

Step 2: Reference

- Effective stress is on page 35-14
- Effective stress is essentially the pressure felt at a point by the soil above it. The unit weight of the soil times the depth.
- Remember to subtract for the buoyant effect of water. Imagine the soil submerged in water and the water is trying to push the soil back up to the surface.
- $\sigma' = \sigma - \mu$
  - $\sigma = \gamma_{soil} H$
  - $\mu = \gamma_{water} H$

Tab CERM

Tab CERM Pg 35-14
Label: Effective Stress

Step 3: Solve for the effective stress

- $\sigma' = \left\{ 74\text{pcf x } 12\text{ft} + 103\text{pcf x } 9\text{ft} + 110\text{pcf x } 6\text{ft} \right\} - \left\{ (9\text{ft}+6\text{ft}) \text{ x} 62.4\text{pcf} \right\}$

$$\boxed{\sigma} \qquad \boxed{\mu}$$

$\sigma' = $ **d) 1539psf**

MIKE HANSEN, P.E.

Problem 24 - Solution

A contractor is purchasing a new track-hoe for $100,000 today. The track-hoe will effectively earn the contractor $40,000 year 1, $32,000 year 2, and $25,000 year 3. If the track-hoe is sold at the end of year 3 for $30,000, what is the loss/gain to the contractor presently (Today's value)? Assume an interest rate of 4%.

a) - $3,547
b) $12,903
c) **$16,942**
d) $27,000

Step 1: Problem Analysis

- Cash Flow
- Calculate the present value of the track-hoe investment.

Step 2: Reference

- Table 86.1 on page 86-7 has the formulas for cash flow
- We are going to use the formula for Future to Present value (F to P) for a single payment.
  - $(1 + i)^{-n}$
  - i, is the interest rate as a decimal. Ex] 4% = 0.04
  - n, is the year of the future value

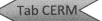

Tab CERM

Tab CERM Pg 86-7
Label: Cash Flow

- The equation is; Present Value $= \dfrac{Future\ Value}{(1+i)^n}$ or $F * (1 + i)^{-n}$, it is the same equation.

Step 3: Solve for the present value

$$\text{Present Value} = \frac{-100,000}{1.04^0} + \overset{-38.4}{\frac{40,000}{1.04^1}} + \overset{-29.5}{\frac{32,000}{1.04^2}} + \overset{-22.2}{\frac{25,000}{1.04^3}} + \overset{-26.6}{\frac{30,000}{1.04^3}}$$

$= $ **c) $16,942**

Problem 25 - Solution

A custom home is being built upon a raised pad. Using the fill heights shown in the figure, how much fill is required? Assume the fill does not require side slopes.

a) 75 yd^3
b) 86 yd^3
c) 111 yd^3
d) **214yd^3**

Step 1: Problem Analysis

- Construction Earthwork
- Calculate the fill required to raise the pad.

Step 2: Reference

- Construction earthwork is in section 79, however simple math techniques can solve this problem
- Calculate the required fill using the area of the pad times the average fill required at each of the corners.
  - V = Area x Fill$_{avg}$ [This is the same with cut or fill]

Step 3: Solve for the required fill volume

- V= (50ft x 60ft) x $\frac{[1.2+ 1.8+ 2.2+ 2.5]}{4}$ = 5775ft^3
- V=5775ft^3 / 27 = **213.89yd^3**

Problem 26 - Solution

A cantilever beam experiences a force acting as an end moment shown in the figure. What is the maximum deflection in the beam?

E=29,000ksi

I=1650in^4

a) -2.50x10^4 in
b) -1.50x10^2 in
c) -0.036 in
d) 0 in

Step 1: Problem Analysis

- Structures – Beam Deflection
- Calculate the maximum deflection in a cantilever beam with end moment

Step 2: Reference

- An excellent reference for this problem is in appendix A-78.
- Case 5: Cantilever with End Moment

  o $\delta_{max} = -\frac{M_0 L^2}{2EI}$

  o M_0 is the moment force – Given, <u>convert to kip-inches</u>

  o L is the length of the beam – Given, but <u>convert to inches</u>

  o E is the modulus of Elasticity - Given

  o I is the moment of inertia - Given

Tab CERM

Tab CERM App. A-78
Label: Beam Deflection

Step 3: Solve for the maximum deflection

- M= $5000 lbf - ft * \dfrac{1\ kip}{1000 lb} * \dfrac{12 in}{1 ft}$ =60 kip-in

- $\delta_{max} = -\dfrac{M_0 L^2}{2EI} = -\dfrac{(60 k-in)(20 ft * 12\frac{in}{ft})^2}{2(29000 ksi)(1650 in^4)}$ = **c) - 0.036 inches**

- Make sure you convert the moment to kip-in and the length to inches

Problem 27 - Solution

A rectangular beam has a 4in x 6in cross section. If the beam experiences a bending moment of 0.7 k-ft, what is the bending stress in the beam?

   a)   0 ksi
   b)   0.12 ksi
   c)   0 .25 ksi
   d)   0 .35 ksi

Step 1: Problem Analysis

   - Structures – Bending Stress
   - Calculate the bending stress in the beam

Step 2: Reference

   - Bending stress is on pg. 44-11

      o $\sigma = \frac{M}{S}$,   eq. 44.38

      o   M is the moment (given), S is the section modulus of the shape

      o $S = \frac{bh^2}{6}$, section modulus for a rectangle, eq. 44.40

Tab CERM

Tab CERM Pg 44-11
Label: Bending Stress

Step 3: Solve for the section modulus of the rectangular cross section

   - $S = \frac{bh^2}{6} = \frac{(4in)6in^2}{6} = 24in^3$

Step 4: Solve for the bending stress

   - $\sigma = \frac{M}{S} = \frac{(0.7k-ft)x12\frac{in}{ft}}{24in^3} =$ **d) 0.35 ksi**

   - Make sure to check your units. Convert your bending moment from feet to inches.

Problem 28 - Solution

Which of the following statements about concrete is true?

a) The hydraulic load on concrete formwork is greatest after the concrete sets.
b) If the water-to-cement ratio of concrete is decreased, water tightness is decreased
c) Adding water to concrete mix increases workability, increases slump, and can decrease the concretes strength.
d) A pound of concrete weighs more than a pound of feathers.

Step 1: Problem Analysis

- Concrete Theory
- Which statement is true?

Step 2: Reference

- The CERM discusses concrete in chapter 49.

Tab CERM Pg 49-1
Label: Concrete

Step 3: Solve for the correct answer

a) The hydraulic load on concrete formwork is greatest after the concrete sets.
    False – The load is greatest just after pouring.
b) If the water-to-cement ratio of concrete is decreased, water tightness is decreased
    False – water tightness increases
c) Adding water to concrete mix increases workability, increases slump, and can decrease the concretes strength.
    True
d) A pound of concrete weighs more than a pound of feathers.
    False – Do not tell me you fell for this me. They weigh the same.

Problem 29 - Solution

The system in the figure to the right
contains an incompressible fluid. What
is the flow rate (Q) of pipe 3?

a) 0.09 ft³/s
b) 0.19 ft³/s
c) 2.65 ft³/s
d) 8.9 ft³/s

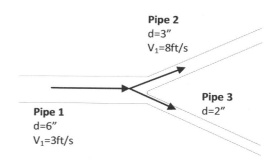

Pipe 2
d=3"
V_1=8ft/s

Pipe 3
d=2"

Pipe 1
d=6"
V_1=3ft/s

Step 1: Problem Analysis

- Hydrology – Continuity Equation
- Calculate the flow rate of Pipe 3

Step 2: Reference

- The continuity equation is on pg 17-2 under section 2: Conservation of
Mass
  - $A_1 V_1 = A_2 V_2 + A_3 V_3 + \dots.$

Tab CERM

Tab CERM Pg 17-2
Label: Continuity E

Step 3: Calculate the pipe areas

- $A1 = \pi \dfrac{d^2}{4} = \pi \dfrac{(0.5 ft)^2}{4} = 0.196 \text{ ft}^2$

- $A2 = \pi \dfrac{d^2}{4} = \pi \dfrac{(0.25) ft^2}{4} = 0.049 \text{ ft}^2$

- $A3 = \pi \dfrac{d^2}{4} = \pi \dfrac{(0.167) ft^2}{4} = 0.022 \text{ ft}^2$

Step 4: Use the continuity equation to calculate v_3

- $A_1 V_1 = A_2 V_2 + A_3 V_3$
- $(0.196 \text{ ft}^2)(3 \text{ ft/s}) = (0.049 \text{ ft}^2)(8 \text{ ft/s}) + (0.022 \text{ ft}^2) V_3$
- $V_3 = 8.9 \text{ ft/s}$

Step 5: Calculate Q_3

- $Q_3 = V_3 A_3 = 8.9 \text{ ft/s } (0.022 \text{ft}^2) =$ **b) 0.19ft³/s**

Problem 30 - Solution

An 8" PVC water system delivers water from a tank (elevation=1890ft) to a subdivision 3000 ft away. If the line is flowing at 2200gpm, what is the head loss from the tank to the subdivision? Assume C=150 for PVC pipe.

a) 25 ft
b) 145 ft
c) **180 ft**
d) 335 ft

Step 1: Problem Analysis

- Hydrology – Hazen Williams
- Whenever you see a "C" value in a problem, you have a Hazen-Williams problem.
- Calculate the head loss in the system.

Step 2: Reference

- Hazen-Williams is on Pg. 17-7

- Head loss (ft) $= \dfrac{10.44 L_{ft} Q_{gpm}^{1.85}}{C^{1.85} d_{inches}^{4.87}}$, Eq. 17.31

- Look at the units. You most likely will not have to convert the units on the PE exam. They will make it so they fit in this format or the one in Eq. 17.30.

Tab CERM

Tab CERM Pg 17-7
Label: Hazen-Williams

Step 3: Solve for head loss

- $h_f = \dfrac{10.44(3000ft)(2200gpm)^{1.85}}{150^{1.85} 8in^{4.87}} = $ **c) 180.09 ft**

Problem 31 - Solution

Rain falls onto the two adjacent areas shown in the figure. Using the rational method, what is the peak flow of the storm water runoff? Assume the runoff flows from Area 1 through Area 2.

a) **9 CFS**
b) 12 CFS
c) 22 CFS
d) 47 CFS

Storm Intensity

time (min)	I (in/hr)
15	1.2
30	2.7
50	3.2

$A_{1=}1.5ac$
$C=0.60$
$t_c=20min$

$A_{2=}2.75ac$
$C=0.70$
$t_c=30min$

Step 1: Problem Analysis

- Hydrology – Rational Method
- Calculate the peak flow of the storm

Step 2: Reference

- The rational method is found on page 20-13
- Q=CIA
    - C = Runoff Coefficient
    - I = Rainfall Intensity
    - A = Total Area

Tab CERM

Tab CERM Pg 20-13
Label: Rational Method

Step 3: Solve for time of concentration and weighted C value

- $t_C = 20+30 = 50min$ – Use this to determine rainfall intensity
- $C=\dfrac{A1(C1)+A2(C2)}{A1+A2}=\dfrac{1.5(0.60)+2.75(0.70)}{1.5+2.75}=0.66$

Step 4: Solve for peak flow

- Q=CIA = 0.66*3.2in/hr*4.25ac = 8.98CFS

Problem 32 - Solution

A concrete cylinder is being tested with a vertical loading. If the concrete breaks at a compressive stress of 4000psi, what was the vertical loading? The diameter of the concrete cylinder is 6 inches.

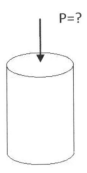

P=?

a) **113kips**
b) 125kips
c) 250kips
d) 300kips

Step 1: Problem Analysis

- Structural – Compressive Strength
- Calculate the axial loading, P

Step 2: Reference

- Pg. 48-4 discusses the compressive strength of concrete.
- f'c= $\sigma = \frac{P}{A}$, eq. 48.1

Tab CERM

Tab CERM Pg48-4 Label: Compressive Strength

Step 3: Solve for P

- A= $\frac{\pi(6in)^2}{4}$ =28.26 in^2
- $P = \sigma A$=4000psi*28.26in^2 = 113,040 lbs or **a) 113kips.**

Problem 33 - Solution

Poisson's ratio v is the ratio of lateral strain to the axial strain. When a sample object is stretched in one direction, it tends to contract somewhere else. Which of the following most aptly applies to Poisson's ratio?

    a) Elastic Strain
    b) Lateral Strain
    c) Axial Strain
    d) Tensile Strength

Step 1: Problem Analysis

- Theory Question – Poisson's Ratio
- Which of the answers is most correct?

Step 2: Reference

- Utilize the index. Poisson's ratio is on 43-4

Tab CERM

Tab CERM Pg 43-4
Label: Poisson's Ratio

Step 3: Solve for the correct answer

- Poisson's ratio is the lateral strain divided by the axial strain. However, the ratio itself is related to an elastic deformation or elastic strain. In section 5: Poisson's Ratio it implicitly denotes that Poisson's ratio applies only to elastic strain.

Problem 34 - Solution

What is the Euler load for the slender vertical column with pinned ends?

$E = 2.9 \times 10^7$ lbf/in^2

$I = 0.50$ in^4

a) **9.9 kips**
b) 19.8 kips
c) 54 kips
d) 117 kips

H = 10ft

Step 1: Problem Analysis

- Euler Load refers to a column bucking theory referred to as Euler Buckling.
- Calculate the Euler load, F_e

Step 2: Reference
- Euler is discussed on pg 45-2 under Slender Columns.
- Eq. 45.1 is used for columns with pinned ends.

  o $F_e = \dfrac{\pi^2 EI}{L^2}$

Tab CERM

Tab CERM Pg 45-2
Label: Euler

Step 3: Solve for the Euler Load

- All information is given so just plug and chug,

  o $F_e = \dfrac{\pi^2 EI}{L^2} = \dfrac{\pi^2 \, 2.9 \times 10^7 (0.5)}{(10ft * 12\frac{in}{ft})^2} =$ **a) 9.9kips**

Problem 35 - Solution

A concrete beam is being designed for use with 2.0 in^2 of steel rebar. The concrete will be a 3000-psi mix (f'$_c$) and the tension steel yield strength (f$_y$) is 60,000 lbf/in^2. What area of concrete (Ac) is required to balance the steel?

a) 24in^2
b) 35in^2
c) 41in^2
d) **47in^2**

Step 1: Problem Analysis

- Reinforced Concrete Beam
- You are looking for the area of concrete (Ac) that is required to balance the steel.

Step 2: Reference

- Section 50-9 describes how to calculate the area of concrete required.

  o $A_c = \dfrac{fyAs}{0.85f'c}$

Step 3: Solve for the area of concrete

- $A_c = \dfrac{(60000psi)(2.0in^2)}{0.85(3000psi)}$ = **d) <u>47.06 in^2</u>**

MIKE HANSEN, P.E.

Problem 36 - Solution

A gravity retaining wall is supporting saturated clay with the properties shown in the figure. What is the <u>total active resultant</u> per unit width of wall?

a) 17 kips/ft
b) 19 kips/ft
c) 20 kips/ft
d) 24 kips/ft

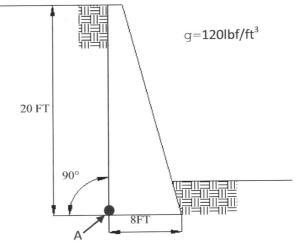

$g = 120lbf/ft^3$

20 FT

90°

8FT

A

Step 1: Problem Analysis

- This problem is an active earth pressure problem.
- We have two theories for active earth problems, Rankine or Coulomb.
- Coulomb theory is used for problems involving friction (d), a sloping backfill (angle b), and an inclined active-side wall face (angle θ).
- Rankine theory disregards wall friction.
- The problem asks to solve for the <u>total active resultant</u> per unit width of wall.

Step 2: Reference

- The equation for total active resultant is found on page 37-4, equation 37.10(b)

Tab CERM

Tab CERM Pg 37-4
Label: Resultant Force

- The formula is $R_a = \frac{1}{2} k_a g H^2$. The only variable we are missing is k_a.
- k_a is solved using the Coulomb or Rankine theory.
- In this problem we have no friction(d), or sloping backfill (angle b).
- Therefore, we use the Rankine theory. Pg 37-4.
- The Rankine formula is $Ka = tan^2(45° - \frac{\phi}{2})$

Tab CERM Pg 37-4
Label: Rankine

- The angle of internal friction ($\emptyset$) in this problem is zero. For saturated clays, the angle of internal friction ($\emptyset$) is zero. There is a note on pg 37-4 that describes this.
- Equation 37.6 is used when we have a sloping backfill (angle b).
- Since there is no sloping backfill in this equation, you can use Eq. 37.7. This equation requires a horizontal backfill and vertical wall.

Step 3: Solve for K_a, equation 37.7

- $Ka = tan^2(45° - \frac{\emptyset}{2})$
- $Ka = tan^2(45° - \frac{0}{2})=1$

Step 4: Solve for R_a

- $R_a = \frac{1}{2} k_a g H^2 = \frac{1}{2} (1)*(120lbf/ft^3)*(20ft)^2 = 24,000lbf/ft$ or **d) 24kips/ft**

The Burm Question:

What if the test asks you to calculate for the overturning moment about point A?

- The overturning moment is the Reactant force (R_a) applied to the height (H/3). It's (H/3) because we have a triangular force acting on the retaining wall.
- So the overturning moment would be $R_a*(H/3) = 24kips/ft * (20ft/3) = 160kips-ft/ft$

Problem 37 - Solution

If the water in the figure were to dry up, which of the following statements is true about the cohesive factor of safety? $\gamma_{sat} = 110$ lb/ft^3, $\gamma_{dry} = 90$ lb/ft^3

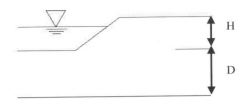

a) The cohesive factor of safety increases.
b) The cohesive factor of safety stays the same.
c) **The cohesive factor of safety decreases.**
d) There is not enough information to make a determination.

Step 1: Problem Analysis

-   This is a geotechnical problem involving Cohesion.
-   Which of the answers is correct about the situation?

Step 2: Reference

-   Cohesion in relation to slope stability is on pg 40-7.
-   $F_{cohesive}$ stands for the cohesive factor of safety.

Tab CERM

Tab CERM Pg 40-7
Label: Cohesion

Step 3: Solve for the correct answer

-   Looking at the equation for the cohesive factor of safety, the γ_{eff} is on the bottom.

    o $F_{cohesive} = \dfrac{N_0 c}{\gamma_{eff} H}$, eq. 40.28

-   $\gamma_{eff} = \gamma_{saturated} - \gamma_{water}$
-   If the soil is no longer saturated from the water, instead of using γ_{eff} we use γ_{dry}. In this case since $\gamma_{dry} > \gamma_{eff}$, **the cohesive factor of safety would decrease.**

Problem 38 - Solution

According to the concrete mix in the table below, what is the total volume of the mix?

a) 0.5 yd^3
b) **1.0 yd^3**
c) 1.75 yd^3
d) 2.25 yd^3

	SSD Weight	Specific Gravity
Low Alkali Cement	442	3.15
Class F Fly Ash	94	2.1
Coarse Aggregate	1809	2.57
Fine Aggregate	1221	2.6
Potable Water	300	1
Total Air	1.50%	-

Step 1: Problem Analysis

- Concrete Volume
- Calculate the volume of Concrete

Step 2: Reference

- No need to reference anything.

$$V = \frac{w}{\gamma} = \frac{w}{S_G \gamma_{water}}$$

Step 3: Solve for the volume of each of the concrete ingredients.

- Once you calculate the volume of each ingredient, add them together, and multiply by the percent of air to get how much air is in the mix.

- Then sum everything together and you have the volume of the mix.

	SSD Weight	Specific Gravity	Volume (ft^3)
Low Alkali Cement	442	3.15	2.25
Class F Fly Ash	94	2.1	0.72
Coarse Aggregate	1809	2.57	11.28
Fine Aggregate	1221	2.6	7.53
Potable Water	300	1	4.81
Total Air	1.50%	-	0.40

26.98ft^3=1.0yd^3

Note: One sack of concrete = 94lbs

Problem 39 - Solution

A soil sample with a volume of 1ft³ was determined to be 50% saturated and 10% porous. If the specific gravity of the soil is 2.10, what is the dry unit weight of the soil?

a) 79 lb/ft³
b) 102 lb/ft³
c) **118 lb/ft³**
d) 167 lb/ft³

Step 1: Problem Analysis

- Soil properties, more specifically dry unit weight.
- Calculate the dry unit weight of the soil.

Step 2: Reference

- Pg 35-9 has a great table of equations relating to soil properties.
- We are looking for the dry density of the soil. $\gamma_d = \dfrac{G_s \gamma_w}{1+e}$
- Since we do not have the void ratio, we need to figure that out.
- $n = \dfrac{e}{1+e}$: n is porosity

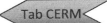

Tab CERM

Tab CERM Pg 35-9
Label: Soil Properties

Step 3: Solve for void ratio then dry density

- $n = \dfrac{e}{1+e}$
- $0.1 = \dfrac{e}{1+e}$, take a guess for 'e'. I tried 0.11. It yields 0.099, pretty close.
  - e=0.11
- Solve for dry unit weight
  - $\gamma_d = \dfrac{G_s \gamma_w}{1+e} = \dfrac{2.10(62.4)}{1+.11} = $ **c) 118.05lbf/ft³**

Problem 40 - Solution

A city with 50,000 people is designing a new wastewater plant to accommodate its current population. The average family is 2.5 persons. If the average amount of solids is 500mg/L, what should the plant be sized for daily intake?

a) 80 lbm/day
b) 2000 lbm/day
c) **26,000 lbm/day**
d) 50,000 lbm/day

Step 1: Problem Analysis

- Wastewater
- Calculate the amount of solids in lbm/day that is generated by the city.

Step 2: Reference

- Pg. 28-2 discusses wastewater quantity. The average volume per person per day is estimated between 100 – 125 gpcd.
- Use the maximum 125 gpcd.

Tab CERM

Tab CERM Pg 28- 2
Label: Wastewater

Step 3: Solve for the daily quantity

- 50,000 x 125gpcd x [500mg/L x 8.34(conversion to lbm/MG)] / 1,000,000
- = **c) 26,063 lbm/day**